Bibliographic information published by the German National Library:

The German National Library lists this publication in the National Bibliography; detailed bibliographic data are available on the Internet at http://dnb.dnb.de .

Imprint:

Copyright © 2010 GRIN Verlag, Open Publishing GmbH
Print and binding: Books on Demand GmbH, Norderstedt Germany
ISBN: 9783668216174

This book at GRIN:

http://www.grin.com/en/e-book/321425/synthesis-characterization-and-single-crystal-structure-of-4-3-methoxy-4-prop-2-yn-1-yloxy-phenyl-2-6-dim

Krunalkumar R. Mehariya, Bhagwati K. Gauni Mehariya

Synthesis, Characterization, and Single Crystal Structure of 4-(3-methoxy-4-(prop-2-yn-1-yloxy)phenyl)-2,6-dimethyl-1,4-dihydropyridine-3,5-dicarbonitrile

GRIN Publishing

GRIN - Your knowledge has value

Since its foundation in 1998, GRIN has specialized in publishing academic texts by students, college teachers and other academics as e-book and printed book. The website www.grin.com is an ideal platform for presenting term papers, final papers, scientific essays, dissertations and specialist books.

Visit us on the internet:

http://www.grin.com/

http://www.facebook.com/grincom

http://www.twitter.com/grin_com

Synthesis, Characterization, and Single Crystal Structure of 4-(3-methoxy-4-(prop-2-yn-1-yloxy)phenyl)-2,6-dimethyl-1,4-dihydropyridine-3,5-dicarbonitrile

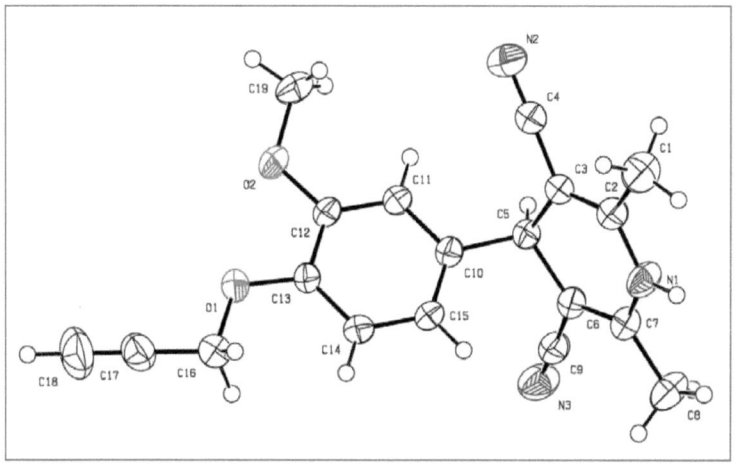

Dr. Krunalkumar R. Mehariya
Mrs. Bhagwati K. Gauni Mehariya

Acknowledgement

Firstly, I would like to thank Prof. Anamik Shah, Department of Chemistry, Saurashtra University for giving me an opportunity to carry out work in his laboratory under his guidance. His wide knowledge in the field of Chemistry research and X-Ray logical way of thinking has been of great value for me to accomplish this chapter work. I am deeply grateful to the Institute for the trust and support that they gave me in order to study in Single Crystal X-Ray Instrumentation, SCX-mini model, Rigaku Company at National Facility for Drug Discovery Complex, Department of Chemistry, Saurashtra University, Rajkot, India.

I owe my special thanks to my beloved wife, Bhawati Gauni Mehariya. My special gratitude is due to my parents, my sisters and their families for their loving support, inspiration to do my best in all matters of life. To them I dedicate this Chapter.

Krunal Mehariya

Contents

1 Introduction

The ability of 1,4-Dihydropyride and its derivatives represent an important class of heterocycles. The synthesis of novel 1,4-Dihydropyride derivatives have gained more importance in recent decades such as, multi drug resistance (mdr) reversal in tumor cell [1-2], antitubercular [3], potential immunomodulating [4] as well. Dihydropyridine core a vital class of Ca^{+2} channel blockers such as Nifedipin and Amlodipine, which are clinically effective in hypertension [5]. 1,4-Dihydropyridines were first successfully synthesized by Hantzsch [6] using a aldehydes, ammonia and ketoester, under reflux in methanol or ethanol which takes longer time with low yields[7-8].

Over the few years, an increasing interest has been persistent on the synthesis of dicyano substitutions on C3 and C5 positions of DHP ring [9-10].

1.1 Experimental

1.1.1 Analysis Protocol

The melting point was determined in an open capillary melting point apparatus. The IR was recorded on Shimadzu FTIR-8400 spectrometer (KBr Pellet method, 400–4000 cm^{-1}). ^{1}H and ^{13}C NMR spectrum was recorded on a Bruker AC 400 MHz using TMS as an internal standard and DMSO-d$_6$ as solvent (chemical shifts are in ppm and coupling constant J in Hz). Mass spectrum was obtained using a Jeol SX 102/DA-6000 spectrometer (for FAB). TLC was performed on 0.25 mm pre-coated silica gel plates (Merck 60 F254) by using ethyl acetate/hexane (1:9) as the eluents.

1.1.2 Reaction Scheme of (3-methoxy-4-(prop-2-yn-1yloxy)phenyl)-2,6-dimethyl-1,4dihydropyridine-3,5-dicarbonitrile

1.1.3 Preparation of (3-methoxy-4-(prop-2-yn-1yloxy)phenyl)-2,6-dimethyl-1,4dihydropyridine-3,5-dicarbonitrile

The present work, reports synthesis and crystal structure of 4-(3-methoxy-4-(prop-2-yn-1yloxy)phenyl)-2,6-dimethyl-1,4dihydropyridine-3,5-dicarbonitrile (3) studied in this work was synthesized from a 3-methoxy-4-(prop-2-yn-1 yloxy) benzaldehyde (1) with (Z)-3-aminobut-2-enenitrile in glacial acetic acid solvent at 60 °C for 30 minute. Reaction was completed as monitored by TLC, the reaction product was filtered to obtain crude product. The precipitate was filtered and recrystallized with ethanol to get pure product. The physical and spectral data of compound are as follow: colorless crystal, yield 90% ; m.p. 180-182 °C.

1.2 Spectral Discussion

1.2.1 IR Spectra

IR spectra of the synthesized compounds were recorded on Shimadzu FT-IR 8400 model using KBr Powder method. Various functional groups present were identified by characteristic frequency obtained for them. Aromatic -CH bond stretching and bending frequencies showed between 3060-3040 cm^{-1} and 1610-1420 cm^{-1} respectively. The –CH bond stretching and bending frequencies for –CH$_3$ and –CH$_2$ group were obtained near 2960-2890 cm^{-1} and 1460-1380 cm^{-1}. The stretching frequency of the characteristic band of secondary N-H group showed in the region of 3500-3210 cm^{-1} with a deformation due to in plane bending at 1660-1570 cm^{-1}. C-O stretching frequency showed at 1230-1140 cm^{-1}. Characteristic frequency of -C≡N bond showed at 2250-2210 cm^{-1}. Characteristic frequency of C-N stretching showed near 1360-1290 cm^{-1}.

1.2.2   Mass Spectra

Mass spectra of the synthesized compounds were recorded on Shimadzu GC-MS-QP- 2010 model using Direct Injection Probe technique. The molecular ion peak was found in agreement with molecular weight of the respective compound.

1.2.3 ^{1}H NMR Spectra

1H NMR spectra of the synthesized compounds were recorded on Bruker Avance II 400 MHz NMR Spectrometer by making a solution of samples in DMSO-d$_6$/CDCl$_3$ solvent using tetramethylsilane (TMS) as the internal standard unless otherwise mentioned. Number of protons identified from 1H NMR spectra and their chemical shift (δ ppm) were in the agreement of the structure of the molecule. J values were calculated to identify o, m and p coupling. In some cases, aromatic protons were obtained as multiplet. Interpretations of representative spectra are discussed as under.

1.2.4 ^{13}C NMR Spectra

^{13}C NMR spectra of the synthesized compounds were recorded on Bruker Avance II 400 MHz NMR Spectrometer by making a solution of samples in DMSO-d$_6$/CDCl$_3$ solvent using tetramethylsilane (TMS) as the internal standard unless otherwise mentioned. Types of carbons identified from NMR spectrum and their chemical shifts (δ ppm) were in the agreement with the structure of the molecule.

1.2.5   Spectral Analysis of (3-methoxy-4-(prop-2-yn-1yloxy)phenyl)-2,6-dimethyl-1,4dihydropyridine-3,5-dicarbonitrile

MS (EI) (m/z): 319; IR (KBr pellet, v; cm^{-1}): 3340 (-NH starching), 3217 (Ar C=C-H starching), 2206 (C≡N starching), 1666 (N-H bend), 1597, 1504, 1427 (Ar C=C starching), 1427 (-CH bending –CH$_2$), 1375 (C-H bend –CH3), 1340 (C-N sec amine vib), 1118 (C-O starching), 864 (-CH overlap); ^{1}H NMR (400 MHz, δ, ppm, DMSO–d$_6$): 9.50 (s, 1H, CH), 7.06-7.04 (d, 1H, J = 8.04 Hz, Ar-H), 6.87 (s, 1H, Ar-H), 6.80-6.78 (d, 1H, j = 7.84, Ar-H), 4.79 (s, 2H, CH$_2$), 4.36 (s, 1H, CH), 3.77 (s, 3H, CH$_3$), 3.41 (s, 1H, NH), 2.04 (s, 6H, 2 ×

CH$_3$); ^{13}C NMR (100 MHz, δ, ppm) 149.11, 146.44, 145.90, 137.75, 119.63, 119.35, 114.13, 111.52, 82.75, 79.33, 78.26, 55.95, 17.73.

1.2.6 Preparation of Single Crystals of (3-methoxy-4-(prop-2-yn-1yloxy)phenyl)-2,6-dimethyl-1,4dihydropyridine-3,5-dicarbonitrile

The Single spot compound of (3-methoxy-4-(prop-2-yn-1yloxy)phenyl)-2,6-dimethyl-1,4dihydropyridine-3,5-dicarbonitrile (500 mg) was taken in 20 mL glacial acetic acid and heated to 60 °C for 10-15 minutes till it dissolved. 200 mg Charcoal was added and further it was heated up to 60 °C for 5-10 minutes. The hot solution was filtered through Celite pad covered by wattmann 41 filter paper. The solution was allowed to cool gradually and kept in a stoppered 100 conical flask slightly opened. The crystals were grown up due to thin layer evaporation.

1.3 X-ray Diffraction Data for (3-methoxy-4-(prop-2-yn-1yloxy)phenyl)-2,6-dimethyl-1,4dihydropyridine-3,5-dicarbonitrile

A colorless block crystal of C$_{19}$H$_{17}$N$_3$O$_2$ having approximate dimensions of 0.550 × 0.300 × 0.300 mm was mounted on a glass fiber. All measurements were made on a Rigaku SCX mini diffractometer using graphite monochromated Mo-Kα radiation. The crystal-to-detector distance was 52.00 mm.

The crystallographic data analysis reveals that crystallizes in the triclinic crystal system, space group, a = 9.1875(8) Å, b = 10.0894(9) Å, c = 10.4209(9) Å, α = 84.555(3)°, β = 73.300(3)°, and γ = 67.740(3)°, Z = 2, V= 856.2(2) Å3, ρcalc= 1.239 g/cm3, space group P-1 (#2). Data Reduction of the 8679 reflections that were collected, 3873 were unique (Rint = 0.0174); equivalent reflections were merged.

Data were collected and processed using Crystal Clear [11]. The linear absorption coefficient, μ, for Mo-Kα radiation is 0.824 cm^{-1}. An empirical absorption correction was applied which resulted in transmission factors ranging from 0.834 to 0.976. The data were corrected for Lorentz and polarization effects.

The structure was solved by direct methods [12] and expanded using Fourier techniques. The non-hydrogen atoms were refined anisotropically. Hydrogen atoms were refined using the riding model. The final cycle of full-matrix least-squares refinement [13] on F_2 was based on 3873 observed reflections and 217 variable parameters and converged. The standard deviation of an observation of unit weight4 was 1.60. Unit weights were used. The maximum and minimum peaks on the final difference Fourier map corresponded to 0.39 and -0.34 e $/Å^3$, respectively.

Neutral atom scattering factors were taken from Cromer and Waber [14]. Anomalous dispersion effects were included in F_{calc} [15]; the values for $\Delta f'$ and $\Delta f''$ were those of Creagh and McAuley [16].

The values for the mass attenuation coefficients are those of Creagh and Hubbell [17]. All calculations were performed using the Crystal Structure [18] crystallographic software package except for refinement, which was performed using SHELXL-97 [19].

1.4   Experimental Details

A. Crystal Data	
Empirical Formula	$C_{19}H_{17}N_3O_2$
Formula Weight	319.36
Crystal Color, Habit	colorless, block
Crystal Dimensions	$0.550 \times 0.300 \times 0.300$ mm
Crystal System	triclinic
Lattice Type	Primitive
Lattice Parameters	a = 9.1875(8) Å
	b = 10.0894(9) Å
	c = 10.4209(9) Å
	α = 84.555(3) ^{o}
	β = 73.300(3) ^{o}
Space Group	P-1 (#2)
Z value	2
D_{calc}	1.239 g/cm3
F_{000}	336.00
$\mu MoK\alpha$)	0.824 cm^{-1}

B. Intensity Measurements	
Diffractometer	SCX mini
Radiation	MoKα (l = 0.71075 Å)
	graphite monochromated
Voltage, Current	50kV, 30mA
Temperature	20.0°C
Detector Aperture	75 mm (diameter)
Data Images	540 exposures
ω oscillation Range	-120.0 - 60.0°
Exposure Rate	10.0 sec./°
Detector Swing Angle	-30.80°
ω oscillation Range	-120.0 - 60.0°
Exposure Rate	10.0 sec./°
Detector Swing Angle	-30.80°
ω oscillation Range	-120.0 - 60.0°

Exposure Rate	10.0 sec./$^\circ$
Detector Swing Angle	-30.80°
Detector Position	52.00 mm
Pixel Size	0.146 mm
$2\theta_{max}$	54.9°
No. of Reflections Measured	Total: 8679 Unique: 3616 (R_{int} = 0.0174)
Corrections	Lorentz-polarization Absorption (trans. factors: 0.834 - 0.976)

C. Structure Solution and Refinement	
Structure Solution	Direct Methods (SIR92)
Refinement	Full-matrix least-squares on F^2
Function Minimized	$\sum w\,(Fo^2 - Fc^2)^2$
Least Squares Weights	$w = 1/\,[\ ^2(Fo^2) + (0.0600 \cdot P)^2$ $+ 0.5337 \cdot\ P\,]$ where $P = (Max(Fo^2,0) + 2Fc^2)/3$
$2\theta_{max}$ cutoff	54.9°
Anomalous Dispersion	All non-hydrogen atoms
No. Observations (All reflections)	3873
No. Variables	217
Reflection/Parameter Ratio	17.85
Residuals:R1 (I>2.00σ(I))	0.0570
Residuals: R (All reflections)	0.0669
Residuals: wR2 (All reflections)	0.2085
Goodness of Fit Indicator	1.603
Max Shift/Error in Final Cycle	0.000
Maximum peak in Final Diff. Map	0.39 $\bar{e}/Å^3$
Minimum peak in Final Diff. Map	$-0.34\ \bar{e}/Å^3$

Table 1. Atomic coordinates and B_{iso}/B_{eq}

Atom	x	y	z	B_{eq}
O1	0.0543(2)	0.3182(2)	0.4597(1)	3.22(3)
O2	0.3715(2)	0.2338(2)	0.4091(2)	4.36(4)
N1	0.2248(2)	0.2228(2)	1.1533(2)	3.47(3)
N2	0.5881(2)	0.3266(2)	0.7634(2)	4.74(4)
N3	0.2277(2)	-0.1837(2)	0.9503(2)	4.44(4)
C1	0.3367(3)	0.4110(2)	1.1083(2)	4.26(4)
C2	0.3193(2)	0.2843(2)	1.0602(2)	2.94(3)
C3	0.3898(2)	0.2288(2)	0.9344(2)	2.74(3)
C4	0.4980(2)	0.2856(2)	0.8413(2)	3.32(4)
C5	0.3591(2)	0.1076(2)	0.8828(2)	2.60(3)
C6	0.2651(2)	0.0449(2)	1.0013(2)	2.70(3)
C7	0.2009(2)	0.1032(2)	1.1262(2)	2.88(3)
C8	0.1044(3)	0.0444(2)	1.2426(2)	4.10(4)
C9	0.2420(2)	-0.0817(2)	0.9754(2)	3.22(4)
C10	0.2709(2)	0.1617(2)	0.7732(2)	2.62(3)
C11	0.3642(2)	0.1656(2)	0.6409(2)	2.88(3)
C12	0.2889(2)	0.2210(2)	0.5395(2)	2.88(3)
C13	0.1180(2)	0.2682(2)	0.5677(2)	2.61(3)
C14	0.0258(2)	0.2643(2)	0.6985(2)	2.86(3)
C15	0.1030(2)	0.2122(2)	0.8005(2)	2.84(3)
C16	-0.1201(2)	0.3737(2)	0.4883(2)	3.50(4)
C17	-0.1618(3)	0.4083(2)	0.3614(2)	3.88(4)
C18	-0.1935(3)	0.4309(3)	0.2585(3)	6.05(6)
C19	0.5340(3)	0.2308(3)	0.3858(2)	5.22(5)

$B_{eq} = 8/3 \ \pi^2(U_{11}(aa^*)^2 + U_{22}(bb^*)^2 + U_{33}(cc^*)^2 + 2U_{12}(aa^*bb^*)\cos g + 2U_{13}(aa^*cc^*)\cos b + 2U_{23}(bb^*cc^*)\cos a)$

Table 2. Atomic coordinates and B_{iso} involving hydrogen atoms

Atom	x	y	Z	B_{iso}	atom	X
H1	0.1785	0.2611	1.2322	4.17	H1	0.1785
H1A	0.2942	0.4930	1.0564	5.12	H1A	0.2942
H1B	0.4501	0.3914	1.0984	5.12	H1B	0.4501
H1C	0.2769	0.4300	1.2010	5.12	H1C	0.2769
H5	0.4652	0.0325	0.8438	3.12	H5	0.4652
H8A	0.1709	-0.0515	1.2596	4.92	H8A	0.1709
H8B	0.0094	0.0435	1.2221	4.92	H8B	0.0094
H8C	0.0712	0.1035	1.3206	4.92	H8C	0.0712
H11	0.4776	0.1307	0.6211	3.46	H11	0.4776
H14	-0.0875	0.2965	0.7183	3.44	H14	-0.0875
H15	0.0401	0.2115	0.8880	3.40	H15	0.0401

H16A	-0.1667	0.4588	0.5448	4.20	H16A	-0.1667
H16B	-0.1633	0.3029	0.5350	4.20	H16B	-0.1633
H18	-0.2187	0.4489	0.1767	7.26	H18	-0.2187
H19A	0.6006	0.1414	0.4154	6.26	H19A	0.6006
H19B	0.5312	0.3082	0.4346	6.26	H19B	0.5312
H19C	0.5791	0.2410	0.2918	6.26	H19C	0.5791

Table 3. Anisotropic displacement parameters					
Atom	U_{11}	U_{22}	U_{33}	U_{12}	U_{13}	U_{23}
O1	0.0372(7)	0.0584(8)	0.0296(6)	-0.0193(6)	-0.0120(5)	0.0031(5)
O2	0.0423(7)	0.099(1)	0.0246(6)	-0.0319(7)	-0.0032(5)	0.0061(6)
N1	0.057(1)	0.0447(8)	0.0294(7)	-0.0254(7)	0.0008(6)	-0.0065(6)
N2	0.058(1)	0.080(2)	0.052(1)	-0.043(1)	-0.0071(8)	0.0066(9)
N3	0.068(1)	0.059(1)	0.049(1)	-0.0390(9)	-0.0015(8)	-0.0085(8)
C1	0.071(2)	0.048(1)	0.048(1)	-0.030(1)	-0.011(1)	-0.0060(8)
C2	0.0377(9)	0.0393(9)	0.0348(9)	-0.0154(7)	-0.0089(7)	0.0009(7)
C3	0.0337(8)	0.0428(9)	0.0315(8)	-0.0179(7)	-0.0101(6)	0.0030(7)
C4	0.0409(9)	0.054(1)	0.0368(9)	-0.0235(8)	-0.0111(7)	0.0019(8)
C5	0.0314(8)	0.0393(8)	0.0277(8)	-0.0131(6)	-0.0072(6)	-0.0001(6)
C6	0.0375(8)	0.0379(8)	0.0290(8)	-0.0158(7)	-0.0097(6)	0.0023(6)
C7	0.0392(9)	0.0379(8)	0.0308(8)	-0.0143(7)	-0.0080(7)	0.0030(6)
C8	0.067(2)	0.051(1)	0.036(1)	-0.030(1)	0.0004(9)	0.0012(8)
C9	0.0429(9)	0.051(1)	0.0309(8)	-0.0225(8)	-0.0058(7)	-0.0004(7)
C10	0.0354(8)	0.0386(8)	0.0271(8)	-0.0160(7)	-0.0075(6)	0.0000(6)
C11	0.0308(8)	0.050(1)	0.0277(8)	-0.0164(7)	-0.0043(6)	-0.0014(7)
C12	0.0363(9)	0.0509(9)	0.0238(8)	-0.0213(7)	-0.0031(6)	-0.0007(7)
C13	0.0371(8)	0.0417(8)	0.0264(8)	-0.0202(7)	-0.0105(6)	0.0022(6)
C14	0.0303(8)	0.0465(9)	0.0324(8)	-0.0170(7)	-0.0051(6)	0.0004(7)
C15	0.0336(8)	0.0471(9)	0.0273(8)	-0.0188(7)	-0.0035(6)	0.0018(7)
C16	0.0395(9)	0.055(1)	0.042(1)	-0.0197(8)	-0.0130(8)	0.0048(8)
C17	0.045(1)	0.053(1)	0.052(1)	-0.0172(8)	-0.0214(9)	0.0043(8)
C18	0.078(2)	0.090(2)	0.066(2)	-0.021(2)	-0.045(2)	0.015(2)
C19	0.044(1)	0.112(2)	0.039(1)	-0.037(2)	0.0006(9)	0.012(1)

The general temperature factor expression: $\exp(-2\pi^2(a^{*2}U_{11}h^2 + b^{*2}U_{22}k^2 + c^{*2}U_{33}l^2 + 2a^*b^*U_{12}hk + 2a^*c^*U_{13}hl + 2b^*c^*U_{23}kl))$

Table 4. Bond lengths (Å)

Atom	Atom	Distance	Atom	Atom	Distance
O1	C13	1.383(2)	O1	C16	1.431(2)
O2	C12	1.3744(19)	O2	C19	1.430(3)
N1	C2	1.375(3)	N1	C7	1.374(3)
N2	C4	1.150(3)	N3	C9	1.149(3)
C1	C2	1.494(4)	C2	C3	1.352(3)
C3	C4	1.425(3)	C3	C5	1.529(3)
C5	C6	1.524(3)	C5	C10	1.529(3)
C6	C7	1.353(3)	C6	C9	1.432(3)
C7	C8	1.501(3)	C10	C11	1.405(2)
C10	C15	1.378(3)	C11	C12	1.386(3)
C12	C13	1.405(3)	C13	C14	1.388(2)
C14	C15	1.398(3)	C16	C17	1.456(3)
C17	C18	1.171(4)			

Table 5. Bond lengths involving hydrogens (Å)

Atom	Atom	Distance	Atom	Atom	Distance
N1	H1	0.860	C1	H1A	0.960
C1	H1B	0.960	C1	H1C	0.960
C5	H5	0.980	C8	H8A	0.960
C8	H8B	0.960	C8	H8C	0.960
C11	H11	0.930	C14	H14	0.930
C15	H15	0.930	C16	H16A	0.970
C16	H16B	0.970	C18	H18	0.930
C19	H19A	0.960	C19	H19B	0.960
C19	H19C	0.960			

Table 6. Bond angles (°)

Atom	Atom	Atom	Angle	Atom	Atom	Atom	Angle
C13	O1	C16	116.30(12)	C12	O2	C19	116.88(16)
C2	N1	C7	123.02(14)	N1	C2	C1	115.85(15)
N1	C2	C3	119.72(18)	C1	C2	C3	124.42(17)
C2	C3	C4	119.53(19)	C2	C3	C5	123.93(16)
C4	C3	C5	116.54(14)	N2	C4	C3	177.4(2)
C3	C5	C6	108.60(13)	C3	C5	C10	110.37(14)
C6	C5	C10	113.03(15)	C5	C6	C7	123.96(17)
C5	C6	C9	116.80(14)	C7	C6	C9	119.22(16)
N1	C7	C6	119.91(16)	N1	C7	C8	115.76(15)
C6	C7	C8	124.32(19)	N3	C9	C6	177.54(17)
C5	C10	C11	118.69(14)	C5	C10	C15	122.36(13)
C11	C10	C15	118.89(16)	C10	C11	C12	120.69(15)
O2	C12	C11	123.97(15)	O2	C12	C13	116.13(15)

C11	C12	C13	119.89(14)	O1	C13	C12	115.77(13)
O1	C13	C14	124.85(15)	C12	C13	C14	119.36(16)
C13	C14	C15	120.12(15)	C10	C15	C14	120.97(14)
O1	C16	C17	107.81(13)	C16	C17	C18	177.6(3)
C13	O1	C16	116.30(12)	C12	O2	C19	116.88(16)
O1	C16	C17	107.81(13)	C16	C17	C18	177.6(3)

Table 7. Bond angles involving hydrogens (°)							
Atom	**Atom**	**Atom**	**Angle**	**Atom**	**Atom**	**Atom**	**Angle**
C2	N1	H1	118.5	C7	N1	H1	118.5
C2	C1	H1A	109.5	C2	C1	H1B	109.5
C2	C1	H1C	109.5	H1A	C1	H1B	109.5
H1A	C1	H1C	109.5	H1B	C1	H1C	109.5
C3	C5	H5	108.2	C6	C5	H5	108.2
C10	C5	H5	108.2	C7	C8	H8A	109.5
C7	C8	H8B	109.5	C7	C8	H8C	109.5
H8A	C8	H8B	109.5	H8A	C8	H8C	109.5
H8B	C8	H8C	109.5	C10	C11	H11	119.7
C12	C11	H11	119.7	C13	C14	H14	119.9
C15	C14	H14	119.9	C10	C15	H15	119.5
C14	C15	H15	119.5	O1	C16	H16A	110.1
O1	C16	H16B	110.1	C17	C16	H16A	110.1
C17	C16	H16B	110.1	H16A	C16	H16B	108.5
C17	C18	H18	180.0	O2	C19	H19A	109.5
O2	C19	H19B	109.5	O2	C19	H19C	109.5
H19A	C19	H19B	109.5	H19A	C19	H19C	109.5
H19B	C19	H19C	109.5				

Table 8. Torsion Angles(^{o})
(Those having bond angles > 160 or < 20 degrees are excluded.)

Atom[1]	Atom[2]	Atom[3]	Atom[4]	Angle
C13	O1	C16	C17	-174.97(14)
C16	O1	C13	C12	-176.79(15)
C16	O1	C13	C14	2.1(3)
C19	O2	C12	C11	-19.3(3)
C19	O2	C12	C13	160.59(18)
C2	N1	C7	C6	-3.4(3)
C2	N1	C7	C8	175.64(13)
C7	N1	C2	C1	-177.91(13)
C7	N1	C2	C3	1.9(3)
N1	C2	C3	C4	-175.02(13)
N1	C2	C3	C5	5.8(3)
C1	C2	C3	C4	4.7(3)
C1	C2	C3	C5	-174.41(15)
C2	C3	C5	C6	-10.3(2)
C2	C3	C5	C10	114.10(16)
C4	C3	C5	C6	170.51(12)
C4	C3	C5	C10	-65.07(16)
C3	C5	C6	C7	8.7(2)
C3	C5	C6	C9	-172.67(12)
C3	C5	C10	C11	84.45(15)
C3	C5	C10	C15	-92.86(18)
C6	C5	C10	C11	-153.71(13)
C6	C5	C10	C15	29.0(2)
C10	C5	C6	C7	-114.09(17)
C10	C5	C6	C9	64.50(16)
C5	C6	C7	N1	-2.7(3)
C5	C6	C7	C8	178.29(13)
C9	C6	C7	N1	178.71(14)
C5	C10	C11	C12	-176.53(15)
C5	C10	C15	C14	178.34(15)
C11	C10	C15	C14	1.0(3)
C15	C10	C11	C12	0.9(3)
C10	C11	C12	O2	177.17(16)
C10	C11	C12	C13	-2.7(3)
O2	C12	C13	O1	1.7(3)
O2	C12	C13	C14	-177.22(15)
C11	C12	C13	O1	-178.34(16)
C11	C12	C13	C14	2.7(3)
O1	C13	C14	C15	-179.68(15)
C12	C13	C14	C15	-0.8(3)
C13	C14	C15	C10	-1.1(3)

Table 9. Intramolecular contacts less than 3.60 Å

Atom	Atom	Distance	Atom	Atom	Distance
O1	O2	2.6116(18)	O1	C18	3.347(4)
N1	C5	2.891(2)	N2	C2	3.465(3)
N2	C5	3.514(3)	N2	C11	3.596(4)
N3	C5	3.521(3)	N3	C7	3.465(3)
C1	C4	2.901(3)	C2	C6	2.798(3)
C2	C10	3.543(3)	C3	C7	2.794(3)
C3	C11	3.274(3)	C3	C15	3.378(3)
C4	C10	3.069(3)	C4	C11	3.229(3)
C6	C15	2.943(3)	C7	C10	3.579(3)
C8	C9	2.898(3)	C9	C10	3.119(3)
C9	C15	3.344(3)	C10	C13	2.804(3)
C11	C14	2.777(3)	C11	C19	2.828(3)
C12	C15	2.778(2)	C14	C16	2.814(3)

Table 10. Intramolecular contacts less than 3.60 Å involving hydrogens

Atom	Atom	Distance	Atom	Atom	Distance
O1	H14	2.663	O2	H11	2.635
N1	H1A	3.047	N1	H1B	3.051
N1	H1C	2.436	N1	H5	3.583
N1	H8A	3.045	N1	H8B	3.059
N1	H8C	2.441	N2	H1A	3.497
N2	H1B	3.389	N2	H5	3.508
N2	H11	3.174	N3	H5	3.536
N3	H8A	3.429	N3	H8B	3.432
C1	H1	2.507	C2	H5	3.180
C3	H1	3.123	C3	H1A	2.779
C3	H1B	2.772	C3	H1C	3.276
C3	H11	3.284	C3	H15	3.452
C4	H1A	2.901	C4	H1B	2.832
C4	H5	2.680	C4	H11	2.983
C5	H11	2.654	C5	H15	2.698
C6	H1	3.124	C6	H8A	2.782
C6	H8B	2.779	C6	H8C	3.281
C6	H15	2.624	C7	H5	3.170
C7	H15	3.148	C8	H1	2.509
C9	H5	2.699	C9	H8A	2.863
C9	H8B	2.861	C9	H15	3.052

C10	H14	3.250	C11	H5	2.604
C11	H15	3.238	C11	H19A	2.660
C11	H19B	2.870	C12	H14	3.252
C12	H19A	2.616	C12	H19B	2.614
C12	H19C	3.189	C13	H11	3.253
C13	H15	3.249	C13	H16A	2.657
C13	H16B	2.592	C14	H16A	2.819
C14	H16B	2.671	C15	H5	3.271
C15	H11	3.235	C16	H14	2.514
C16	H18	3.556	C18	H16A	3.111
C18	H16B	3.087	C19	H11	2.549
H1	H1A	3.149	H1	H1B	3.168
H1	H1C	2.173	H1	H8A	3.165
H1	H8B	3.157	H1	H8C	2.177
H11	H19C	3.479	H14	H15	2.320
H14	H16A	2.402	H14	H16B	2.201

Table 11. Intermolecular contacts less than 3.60 Å					
Atom	Atom	Distance	Atom	Atom	Distance
O1	O1^1	3.5346(19)	O1	N1^2	3.1841(18)
O1	C8^2	3.548(3)	O1	C16^1	3.501(3)
O2	N1^2	3.342(3)	O2	C1^2	3.493(3)
N1	O1^3	3.1841(18)	N1	O2^3	3.342(3)
N1	C17^3	3.482(2)	N1	C18^3	3.504(3)
N2	C1^4	3.445(4)	N2	C16^5	3.453(3)
N2	C18^1	3.582(3)	N3	C4^6	3.589(3)
N3	C15^7	3.470(3)	N3	C19^8	3.522(3)
C1	O2^3	3.493(3)	C1	N2^4	3.445(4)
C3	C6^6	3.517(2)	C3	C9^6	3.506(3)
C4	N3^6	3.589(3)	C4	C9^6	3.427(3)
C4	C18^1	3.484(3)	C6	C3^6	3.517(2)
C8	O1^3	3.548(3)	C9	C3^6	3.506(3)
C9	C4^6	3.427(3)	C15	N3^7	3.470(3)
C16	O1^1	3.501(3)	C16	N2^9	3.453(3)
C17	N1^2	3.482(2)	C18	N1^2	3.504(3)
C18	N2^1	3.582(3)	C18	C4^1	3.484(3)
C19	N3^8	3.522(3)			

Symmetry Operators:

(1) -X,-Y+1,-Z+1   (2) X,Y,Z-1

(3) X,Y,Z+1   (4) -X+1,-Y+1,-Z+2

(5) X+1,Y,Z   (6) -X+1,-Y,-Z+2

(7) -X,-Y,-Z+2   (8) -X+1,-Y,-Z+1

(9) X-1,Y,Z

Table 12. Intermolecular contacts less than 3.60 Å involving hydrogens						
Atom	**Atom**	**Distance**		**Atom**	**Atom**	**Distance**
O1	H1^1	2.337		O1	H1C^1	3.275
O1	H8C^1	2.651		O1	H16A^2	2.800
O2	H1^1	2.835		O2	H1B^1	3.475
O2	H1C^1	2.888		O2	H5^3	3.525
O2	H11^3	3.413		O2	H16A^2	2.955
N1	H5^4	3.038		N2	H1A^5	3.364
N2	H1B^5	3.166		N2	H1C^5	3.234
N2	H8A^4	2.798		N2	H14^6	2.781
N2	H16A^6	3.286		N2	H16B^6	2.740
N2	H18^2	3.204		N3	H1A^7	3.229
N3	H1B^4	2.837		N3	H8B^8	3.066
N3	H14^8	3.563		N3	H15^8	2.650
N3	H18^9	3.133		N3	H19C^3	2.603
C1	H1B^5	3.489		C1	H14^{10}	3.276
C1	H18^2	3.448		C1	H19C^{11}	3.238
C2	H5^4	3.298		C2	H18^2	3.516
C3	H5^4	3.468		C3	H18^2	3.293
C4	H8A^4	3.021		C4	H18^2	2.962
C5	H5^4	3.588		C6	H5^4	3.155
C7	H5^4	2.945		C8	H5^4	3.580
C8	H14^8	3.482		C9	H1B^4	3.297
C9	H8B^8	3.409		C9	H15^8	3.249
C9	H19C^3	3.042		C10	H19A^3	3.417
C11	H19A^3	3.085		C12	H1^1	3.573
C12	H16A^2	3.123		C12	H19A^3	3.428
C13	H1^1	3.383		C13	H8C^1	3.401
C13	H16A^2	3.030		C14	H1A^{10}	3.547
C14	H1C^{10}	3.284		C14	H8A^8	3.221
C14	H8B^8	3.249		C15	H8B^8	3.169

C16	H1^1	3.154		C16	H1C^{10}	3.586
C16	H8C^1	3.014		C16	H16A^2	3.558
C17	H1^1	2.870		C17	H8C^1	2.993
C17	H19B^{12}	3.214		C18	H1^1	3.131
C18	H1A^2	3.597		C18	H8C^1	3.414
C18	H19B^{12}	3.251		C18	H19C^{12}	3.263
C19	H1B^1	3.412		C19	H1C^1	3.407
C19	H16A^2	3.569		H1	O1^{11}	2.337
H1	O2^{11}	2.835		H1	C12^{11}	3.573
H1	C13^{11}	3.383		H1	C16^{11}	3.154
H1	C17^{11}	2.870		H1	C18^{11}	3.131
H1	H5^4	3.425		H1A	N2^5	3.364
H1A	N3^{13}	3.229		H1A	C14^{10}	3.547
H1A	C18^2	3.597		H1A	H1B^5	3.003
H1A	H14^{10}	2.993		H1A	H15^{10}	3.321
H1A	H18^2	2.667		H1B	O2^{11}	3.475
H1B	N2^5	3.166		H1B	N3^4	2.837
H1B	C1^5	3.489		H1B	C9^4	3.297
H1B	C19^{11}	3.412		H1B	H1A^5	3.003
H1B	H1B^5	3.073		H1B	H19C^{11}	2.682
H1C	O1^{11}	3.275		H1C	O2^{11}	2.888
H1C	N2^5	3.234		H1C	C14^{10}	3.284
H1C	C16^{10}	3.586		H1C	C19^{11}	3.407
H1C	H14^{10}	2.704		H1C	H16A^{10}	2.732
H1C	H19C^{11}	3.065		H5	O2^3	3.525
H5	N1^4	3.038		H5	C2^4	3.298
H5	C3^4	3.468		H5	C5^4	3.588
H5	C6^4	3.155		H5	C7^4	2.945
H5	C8^4	3.580		H5	H1^4	3.425
H5	H5^4	3.452		H5	H8A^4	3.280
H5	H19C^3	3.419		H8A	N2^4	2.798
H8A	C4^4	3.021		H8A	C14^8	3.221
H8A	H5^4	3.280		H8A	H11^4	3.575
H8A	H14^8	2.824		H8A	H16B^8	3.169
H8B	N3^8	3.066		H8B	C9^8	3.409
H8B	C14^8	3.249		H8B	C15^8	3.169
H8B	H14^8	3.264		H8B	H15^8	3.144
H8B	H19A^{14}	3.515		H8B	H19C^{14}	3.575
H8C	O1^{11}	2.651		H8C	C13^{11}	3.401
H8C	C16^{11}	3.014		H8C	C17^{11}	2.993
H8C	C18^{11}	3.414		H8C	H16B^{11}	2.913
H11	O2^3	3.413		H11	H8A^4	3.575
H11	H19A^3	3.169		H14	N2^{12}	2.781
H14	N3^8	3.563		H14	C1^{10}	3.276
H14	C8^8	3.482		H14	H1A^{10}	2.993
H14	H1C^{10}	2.704		H14	H8A^8	2.824

H14	H8B[8]	3.264		H15	N3[8]	2.650
H15	C9[8]	3.249		H15	H1A[10]	3.321
H15	H8B[8]	3.144		H16A	O1[2]	2.800
H16A	O2[2]	2.955		H16A	N2[12]	3.286
H16A	C12[2]	3.123		H16A	C13[2]	3.030
H16A	C16[2]	3.558		H16A	C19[2]	3.569
H16A	H1C[10]	2.732		H16A	H16A[2]	3.328
H16A	H19B[2]	3.241		H16B	N2[12]	2.740
H16B	H8A[8]	3.169		H16B	H8C[1]	2.913
H16B	H19B[12]	3.247		H18	N2[2]	3.204
H18	N3[9]	3.133		H18	C1[2]	3.448
H18	C2[2]	3.516		H18	C3[2]	3.293
H18	C4[2]	2.962		H18	H1A[2]	2.667
H18	H19B[12]	3.565		H18	H19C[12]	3.238
H19A	C10[3]	3.417		H19A	C11[3]	3.085
H19A	C12[3]	3.428		H19A	H8B[15]	3.515
H19A	H11[3]	3.169		H19B	C17[6]	3.214
H19B	C18[6]	3.251		H19B	H16A[2]	3.241
H19B	H16B[6]	3.247		H19B	H18[6]	3.565
H19C	N3[3]	2.603		H19C	C1[1]	3.238
H19C	C9[3]	3.042		H19C	C18[6]	3.263
H19C	H1B[1]	2.682		H19C	H1C[1]	3.065
H19C	H5[3]	3.419		H19C	H8B[15]	3.575
H19C	H18[6]	3.238				

Symmetry Operators:

(1) X,Y,Z-1

(2) -X,-Y+1,-Z+1

(3) -X+1,-Y,-Z+1

(4) -X+1,-Y,-Z+2

(5) -X+1,-Y+1,-Z+2

(6) X+1,Y,Z

(7) X,Y-1,Z

(8) -X,-Y,-Z+2

(9) -X,-Y,-Z+1

(10) -X,-Y+1,-Z+2

(11) X,Y,Z+1

(12) X-1,Y,Z

(13) X,Y+1,Z

(14) X-1,Y,Z+1

(15) X+1,Y,Z-1

1.5 Crystal Structure Images

1.5.1 Represents the ORTEP of the molecule (3) with thermal ellipsoids drawn at 50% probability

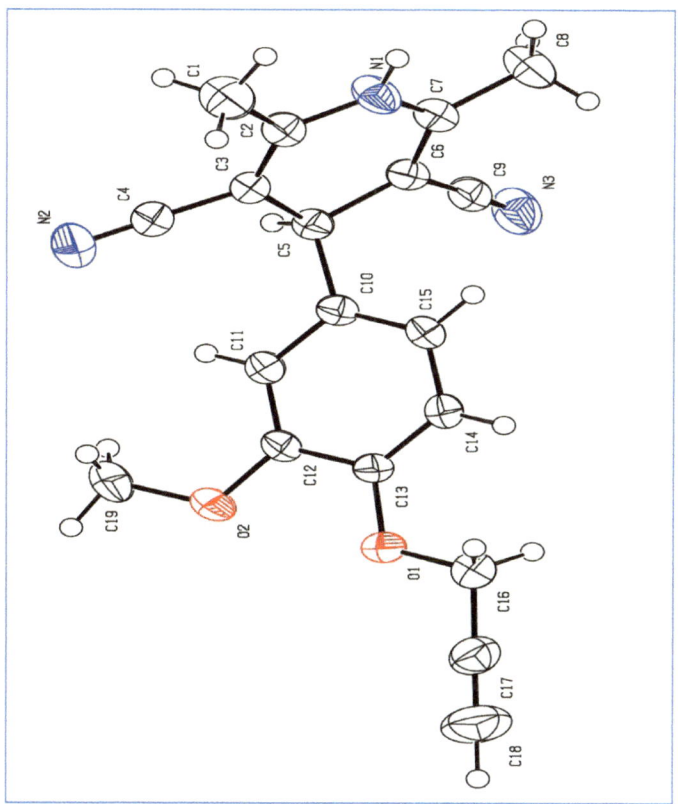

1.6 Packing diagram of the molecules when viewed down the b axis.

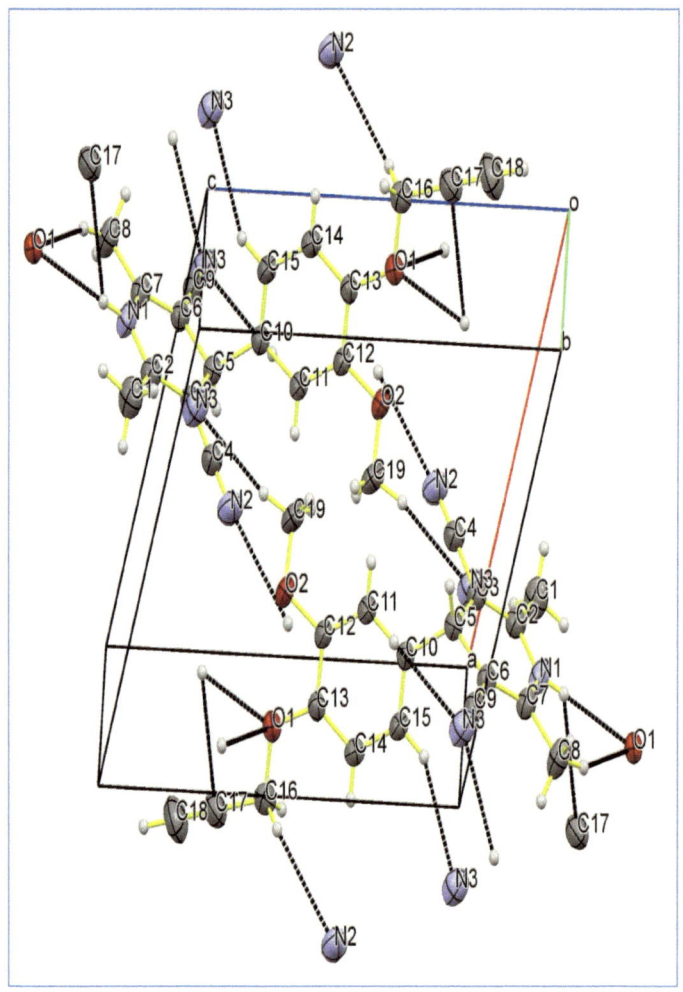

1.7 Representative spectrums

1.7.1 Infrared Spectrum of Compound (3)

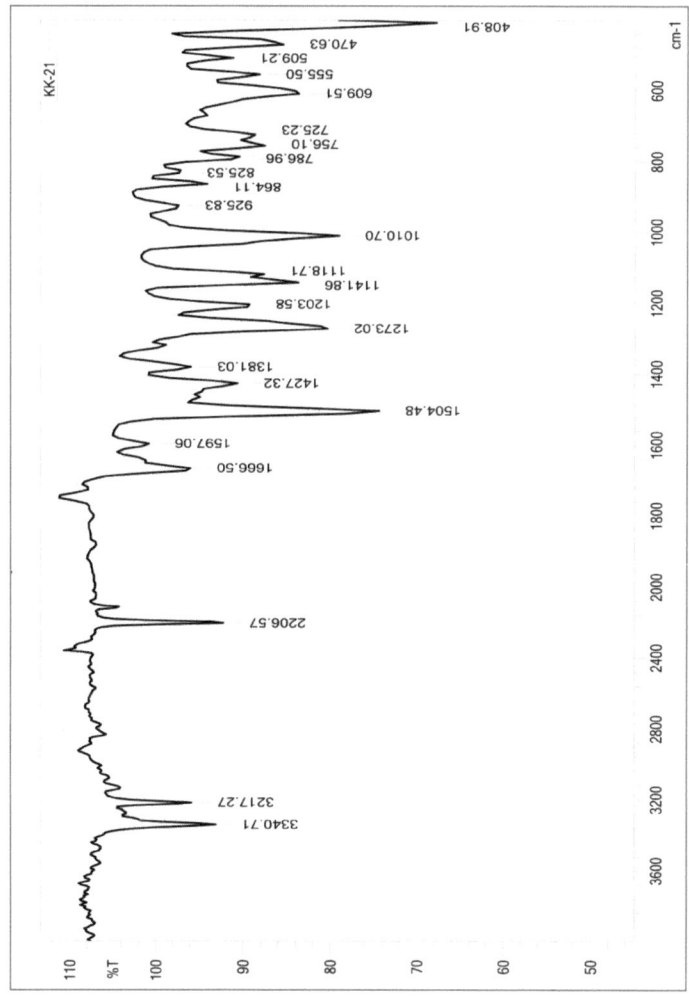

1.7.2 Mass Spectrum of Compound (3)

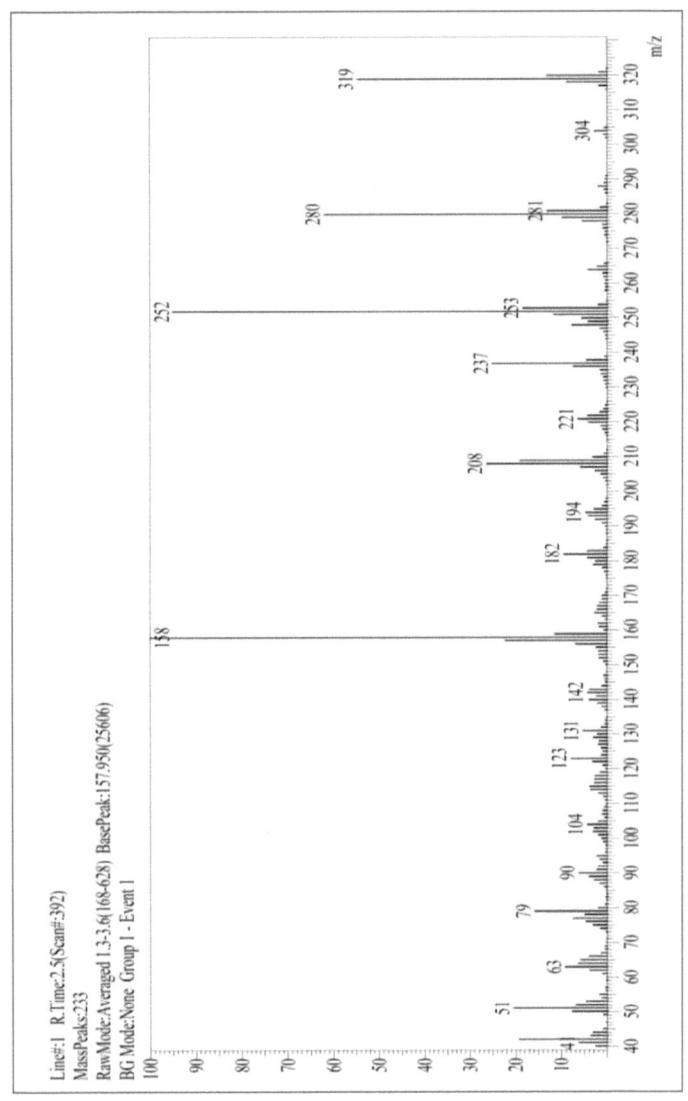

Line#:1 R.Time:2.5(Scan#:392)
MassPeaks:233
RawMode:Averaged 1.3-3.6(168-628) BasePeak:157.95(0.25606)
BG Mode:None Group 1 - Event 1

1.7.3 ¹H Spectrum of Compound (3)

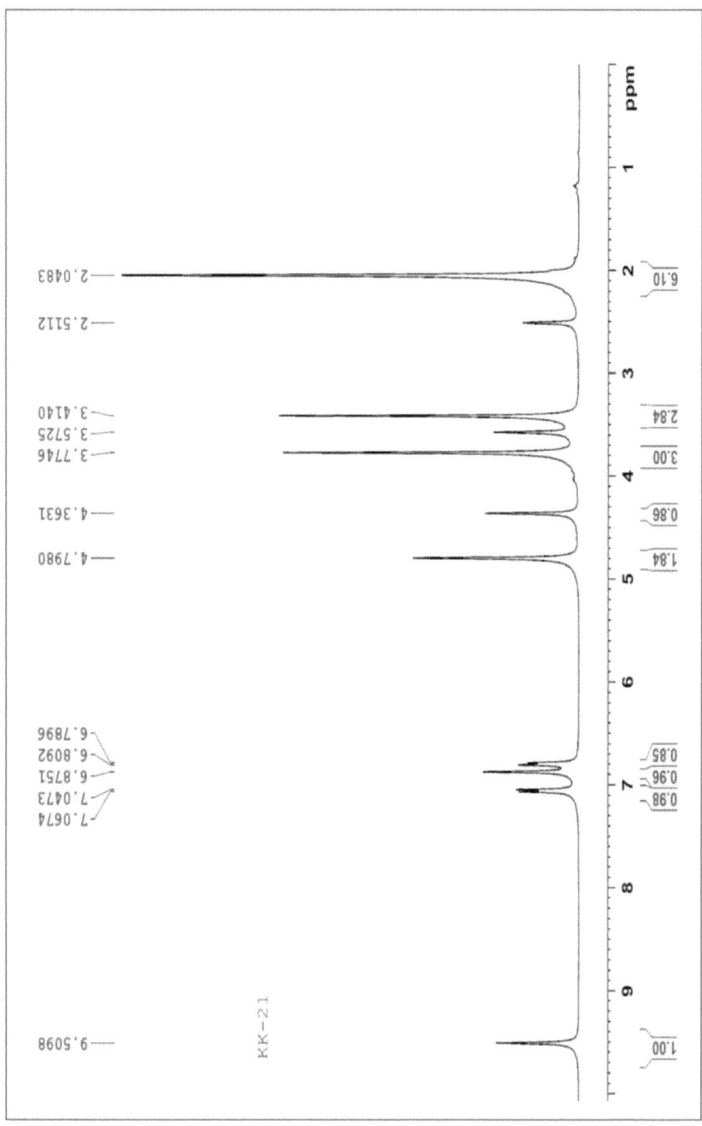

KK-21

2.0483
2.5112
3.4140
3.5725
3.7746
4.3631
4.7980
6.7896
6.8092
6.8751
7.0473
7.0674
9.5098

6.10
2.84
3.00
0.86
1.84
0.85
0.96
0.98
1.00

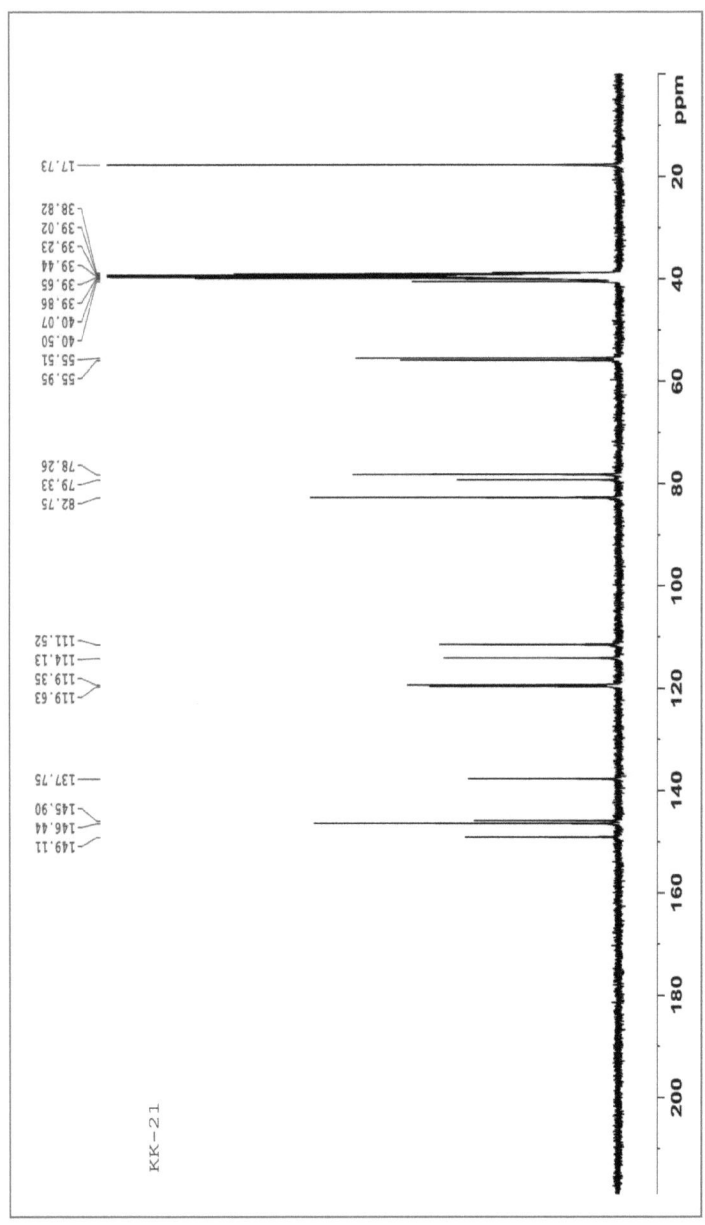

1.8 References

1. (a) Triggle,; Lang, D. A.; Janis, R. A. Med. Res. Rev. 1989, 9, 123. (b). Goldmann, S.; Stoltefuss, J.; Angew. Chem. Int. Ed. Engl. 1991, 30, 1559.

2. (a) Siva Lakshmi Devi,Y.; Srinivasa Rao, M.; Satish, G.; Jyothi, K.; BabuRao, T.; Omdutt, Magn. Reson. Chem. 2007, 45, 688. (b) Gotrane, D.; Deshmukh, R.; Ranade, P.; Sonawane, S.; Bhawal, B.; Gharpure, M.; Gurjar, M. Org. Process Res. Dev. 2010, 14, 640. (e) Connie, M.; Andrea, G.; Paula, R. Annals of Neurology. 2012, 71(3), 362-369.

3. (a) Stout, D. M.; Meyers, A. I.; Recent advances in the chemistry of dihydropyridines. Chem. Rev. 1982, 82, 223-243. (b) Eisner, U.; Kuthan, J.; The chemistry of hydropyridines. Chem. Rev. 1972, 72, 1-42.

4. (a) Loev, B.; Snader, K. M. J. Org. Chem. 1965, 30, 1914. (b) Alajarin, R.; Vaquero, J. J.; Garcia, J. L. N.; J. Alvarez-Builla, Synlett. 1992, 297.

5. Sausins, A.; Duburs, G. Heterocycles. 1988, 27, 269

6. Ilavsky, D. M.; Milata, V.; Synthesis and spectral properties of unsymmetrically 3,5-disubstituted 2,6- dimethyl-1,4-dihydropyridiens. Collect. Czech. Chem. Comm. 1996, 61, 1233-1243.

7. Zenouz, A. M.; Oskuie, M. R.; Mollazadeh, S.; Synthesis of novel asymmetrical 1,4-dihydropyridine derivatives. Synth. Commun. 2005, 35(22), 2895-2903.

8. Zenouz, A. M.; Oskuie, M. R.; Mollazadeh, S.; Synthesis of novel asymmetrical 1,4-dihydropyridine derivatives. Synth. Commun. 2005, 35(22), 2895-2903.

9. Zenouz, A. M.; Allahverdi, S. S.; Raissossadat, M.; Sadeghi, Q. Synthesis of the C-2 functionalized 1,4- dihydropyridines. Asian J. Chem. 2005, 17(4), 2639- 2643.

10. Shah, A. K.; Chemistry & Biology Interface, 2012, 2, 4, 220-227

11. CrystalClear: Rigaku Corporation, 1999. CrystalClear Software User's Guide, Molecular Structure Corporation,(c) 2000.J.W.Pflugrath (1999) Acta Cryst. D55, 1718-1725.

12. SIR92: Altomare, A., Cascarano, G., Giacovazzo, C., Guagliardi, A., Burla, M., Polidori, G., and Camalli, M. (1994) J. Appl. Cryst., 27, 435.

13. Least Squares function minimized: (SHELXL97)
$Sw(F_o^2-F_c^2)^2$ where w = Least Squares weights.

14. Standard deviation of an observation of unit weight:
$[Sw(F_o^2-F_c^2)^2/(N_o-N_v)]^{1/2}$
Where: N_o = number of observations, N_v = number of variables

15. Cromer, D. T. & Waber, J. T.; "International Tables for X-ray Crystallography", Vol. IV, The Kynoch Press, Birmingham, England, Table 2.2 A (1974).

16. Ibers, J. A. & Hamilton, W. C.; Acta Crystallogr., 17, 781 (1964).

17. Creagh, D. C. & McAuley, W.J .; "International Tables for Crystallography", Vol C, (A.J.C. Wilson, ed.), Kluwer Academic Publishers, Boston, Table 4.2.6.8, pages 219-222 (1992).

18. Creagh, D. C. & Hubbell, J.H..; "International Tables for Crystallography", Vol C, (A.J.C. Wilson, ed.), Kluwer Academic Publishers, Boston, Table 4.2.4.3, pages 200-206 (1992).

19. CrystalStructure 4.0: Crystal Structure Analysis Package, Rigaku Corporation (2000-2010). Tokyo 196-8666, Japan.

20. SHELX97: Sheldrick, G.M. (2008). Acta Cryst. A64, 112-122.